To Rosema

THE KEY TO THE UNIVERSE

By
Gladys L. Pohl

Best regards
Gladys L. Pohl.

j.pohl associates

Published By

J. POHL ASSOCIATES
461 Spring Run Road
Coraopolis, PA 15108

ISBN 0-939332-10-8

Cover Design by T. W. McKenzie

Dedicated To

My favorite librarian, my husband Bob.
Dave and Don for their help.
My loving family and friends.
C, B and M
for their patience,
encouragement
and support
in my quest for truth

ACKNOWLEDGEMENT

A personal thanks to my young protege, Linda Zimmerman, a chemist with an inquisitive mind and a love for all science. Her help in the final preparation of this book was greatly appreciated.

ABOUT THE AUTHOR

Pioneers are a rare breed. They require a level of ingenuity and determination that only a handful of people possess. Gladys Pohl is one of those people. She is a pioneer of the universe and her frontier stretches from the distant quasars to the close confines of sub-atomic particles.

She sought to blaze a trail of reason through the confused world of scientific dogma. This book is the result of that search and it is the work of a mind that never stops thinking, probing and rethinking. The vitality with which she approaches her work is refreshing and inspiring. Nothing diminishes her desire for truth, even when that truth seems unobtainable.

It was a privilege to have worked with Gladys. It was an experience that was stimulating and at times a bit overwhelming, just exactly what you would expect from a pioneer.

Linda Zimmerman

PREFACE

It is my desire that my theory, *The Key to the Universe,* as presented in this book, will give a better understanding of cosmic law and correct man's misconceptions of physical reality.

It is said that Dr. Albert Einstein gazed at the sky and remarked that there has to be a great universal law that governs all things within the confines of space or we could have no lasting science. That remark has plagued my inquisitive mind for 25 years. This has to be the Law of Imbalance, for without imbalance we would have no universe. I trust that this is the law that scientists have been searching for, because I have climbed the steps of their written knowledge to reach this conclusion.

$E=mc^2$ holds up well in this study. When energy exceeds mass which brings about an imbalance, the mass divides often into a chain reaction. Otto Hahn and Lisa Meitner drew the conclusion that the nucleus split in half, even though it is impossible to photograph. This conclusion may prove to be wrong. For the atom did not split in half but rather divided and went into a chain reaction, eventually breaking down to minute particles easily disbursed or spent.

The simplistic nature of this Law of Imbalance was as

much of a shock to me as it will be for anyone who has attempted to find an acceptable answer to the cosmic riddle. Perhaps it is because the answer is so simple that until now it has defied all solution.

Gladys L. Pohl
September 13, 1984

THE MAKING OF A LOCKSMITH

Even to the eyes of the most casual observer, the universe is a spectacular place. The myriad of sights and sounds are enough to boggle the most sophisticated mind. From the tiniest sub-atomic particle, to the greatest galactic cluster, ours is a world of fascination and wonder.

However, we are more than just observers in this awe-inspiring universe. We are made of the same elements and subject to the same endless cycle of creation and destruction. Nothing acts solely on its own. The full extent of our interdependence may never be known, but each new discovery seems to tighten the threads of the cosmic cloth.

The ultimate question then is: Is there an all-embracing loom which holds this cloth, or were we cut and sent adrift in the beginning? Did the shuttle of creation weave its patterns without purpose or direction, or does some great scheme guide every strand?

These are questions whose time has come. Now, as in no other period of history, man is prepared to discover the truth in relation to every level of nature. We have lived too long in darkness and ignorance. Every day, science adds a new piece to the puzzle, yet few scientists have taken a step back to see the whole structure developing. Perhaps the most famous man who did so was Dr. Albert Einstein.

Dr. Einstein felt very strongly about the existence of a Unified Field Theory and spent many years of his life searching for it. "The idea that there are two structures of space independent of each other, the metric-gravitational and the electro-magnetic, is intolerable to the theoretical spirit." He felt that no one should be content to tolerate such a disjointed universe.

I have always greatly admired Dr. Albert Einstein. His tireless pursuit of the truth inspired me in my own quest for knowledge. I kept Einstein's unanswered question about a Unified Field Theory in my mind as I, too, sought a unifying law — a "key" to the universe which would unlock the secrets that countless generations of scientists have sought.

The events and phenomena I studied are not new. What is new is the manner in which I viewed them and the conclusions I drew.

Some may be shocked by the simplicity of this "key" but I have yet to uncover any contradictions, even after repeatedly testing my theory against each new fact revealed by science.

My purpose in writing this book was to aid science in its quest for truth and to expose this unifying law from which all things spring.

There are so many signs which point the way to cosmic harmony, not cosmic chaos. From the graceful structure of DNA to the orderly movement of the planets, there are enough examples to draw the conclusion that a common force is propelling everything. It is this universal propellant which is the basis for the unifying law and I will follow it down the many paths which my research leads me.

I have often been asked what credentials I possess. My response has always been, "Thinking and creativity are the privilege of everyone." I have a great love of science and the energy and sense to think things through. The "key" and its implications are too great to ignore and should be dealt with honestly and with an open mind. If my "key" is truly *The Key to the Universe,* there should be no delay in employing it. Albert would have wanted it that way.

THE SEARCH BEGINS

Anyone who is acquainted with a victim of cancer is naturally curious about the cause of the cancer and the latest treatment available. When it is you, yourself, who are stricken, the curiosity becomes a passion for answers which must be found. Cancer was widespread in my family, and in 1959 it was thought that I, too, had cancer.

I had dozens of questions about my condition and I turned to the nurses and doctors at the hospital for help. Much to my surprise, the most frequent response to my questions was, "I don't know." There I was being poked, prodded and cut open for something the medical world did not understand. I was disillusioned and appalled, especially after surgery, which turned out to be a false alarm.

Why did cancer cells reproduce so widely? What triggered this peculiar behavior? These were the two questions which bothered me the most. I tried to see cancer as being subject to a natural law. I sought parallels with other types of reactions, I reasoned that it probably wasn't a single bacteria or virus at work, but a by-product of their action which was the culprit. The fact that non-biological agents, such as radiation, were also known to initiate cancer formation further backed up my assumption. There had to be a common force at work in both the living and non-living promoters of this disease.

The problem was where to begin. I read journals, visited hospitals and spoke with professors. I tried in vain to find reasonable theories as to the nature of cancer. There was no consensus of opinion among the experts.

Unfortunately for the patients, cancer was treated as a foreign substance. Often this isolated treatment was carried out with a total disregard for the rest of the patient. Both before and after surgery I was painfully aware of that fact. My hope was that in discovering what made a tumor grow I could simultaneously find a more humane way of eradicating it. When I had recovered completely, I focused all of my energies on this quest. I never imagined how far it would take me, or that it would begin with something as common as an egg.

THE KEY IS HATCHED

In order to understand more about how cells function, I decided to do some experiments with hen eggs. Having grown up on a farm, I knew that eggs had to be incubated for twenty-one days at a temperature of 99.5-100° F. These are facts that every farmer knows from experience. Naturally, I would not take them for granted until I proved to myself that there were no other conditions under which an egg would hatch. With several experiments in mind, I made my way to a local poultry farm.

The owner of the farm was able to supply me with both fertile and non-fertile eggs. My first experiment was to try and hatch eggs at a higher temperature, e.g. 102° F, in hopes of hatching the eggs in eighteen days instead of the usual twenty-one. This resulted in half-cooked, rotten eggs. By increasing the temperature, I had killed the cells. Next, by enlisting the aid of my dentist, I was able to subject ten of the fertile eggs to a dose of X-rays. After twenty-one days, two normal chicks were hatched, seven were mutations and one egg had rotted. Radiation didn't affect the conditions of hatching, just the end result. I had changed the pattern of life.

My supply of eggs ran out and I returned to the poultry farm for some more. I asked to see the hens actually laying but was told that this area is out of bounds. The reason visitors are barred is because a frightened hen produces an egg with a large blood spot on the nucleus. These eggs cannot be sold. The owner explained that a five minute visit would cost him a few dozen eggs. This was exciting news. I took my fresh supply of eggs and headed for the hardware store. I wanted to try to duplicate the damage inflicted on the egg by a frightened hen.

I bought a large flashlight battery and some wire and headed home. As I was arranging the battery, I realized that I had left the eggs in the car. It was the first of November and it was quite cold. The eggs, too, had gotten cold, but I was anxious to try my make-shift shock inducer. I made a slight crack in three eggs and, because the egg shell is a natural insulator, applied the wires to the exposed areas. After half an hour, I opened the eggs and discovered the same blood spots the poultry farm owner had described. By this time the other eggs had warmed up and I applied the current to them.

To my surprise, the current had no effect on the warm eggs. There were no blood spots and everything appeared normal. Apparently electricity did not flow as well in a warm egg as it did in an abnormally cold one. These results in themselves were not conclusive, but they did start me thinking about the importance of temperature in the development of a chick. I felt as if I was on the verge of some important discoveries.

I thought back to the old Tennessee phrase, "crabby as an old setting hen." Setting hens always ran a slight fever which seemed to make them irritable. I knew from my nursing days that pregnant women also usually run a low grade fever. Obviously, temperature and birth are closely related, but I still felt as if I were missing the real point.

Why should a few degrees alter the cycle of development so much?

The following spring, a mother robin and a mother mockingbird built their nests in my backyard. It was unusually hot that year and I noticed the two birds going through some peculiar movements during the period that they were sitting on their eggs. In the heat of midday, they flew to the ground and hopped around with their wings spread out. I couldn't explain this seemingly irrational behavior until I remembered that eggs need a closely regulated temperature in order to develop properly. The clever birds were fanning themselves to bring their body temperatures down in order to protect their eggs. They clearly recognized that heat energy was a critical factor. Had I been as smart as those birds, my hot pursuit of answers would have ended long ago.

As I looked at my egg experiments and remembered the robin and mockingbird, things began to make sense. DNA may contain the pattern of life, but it is a worthless blueprint without the energy to set it in motion. Normal body cells have all the necessary information to reproduce and yet they do not divide without some outside stimulus. Cancer arises from previously stable cells and begins to multiply rapidly and uncontrollably. In all these instances I saw one common thread — the scales of equilibrium were being tipped.

The last experiment and the one that led to the breakthrough was when I subjected an egg to a 24-hour hook-up with the battery. The result of this experiment was that I had six small yokes instead of one. The conclusion is that energy is definitely the trigger.

Everything became clear. Years of work suddenly fell into place. The driving force behind all the works of nature, from the creation of the universe to the spread of cancer, is nothing more than a reaction to an imbalance of energies. When a cell is thrown off balance, a force is created that triggers that cell to reproduce. All the genetic material in the world cannot produce a single thing without the proper amount of energy being applied to it.

Another example that illustrates this law of imbalance can be found in an ordinary garden. It is a marvel in itself that a handful of seeds can be turned into bushels of fruits and vegetables. However, all the fertilizer and water is useless unless enough warmth is present to germinate the seeds. Each seed is a tiny blueprint containing all the information necessary to produce a vigorous plant. Yet it is only so much food for the worms if the sun doesn't make an appearance. It is energy applied to the blueprint which sets the cycle in motion.

This is The Key to the Universe. Nothing exists where there is true and lasting equilibrium. Every function, every reaction depends upon an initial imbalance of energies. If every force in the universe remained in a state of equilibrium, there would be no universe as we know it. It is as simple as that!

Simplicity, and the pursuit of it, is the way of all things. All that exists follows the simple path of evolution, and evolution is nothing more than nature striving for a perfect, balanced state. If we were able to see beneath the inner workings of the cosmos, we would see a system working toward ever-increasing harmony. When that harmony is absolute, the universe will indeed be in a state of equilibrium and evolution will cease to exist. At this point, however, our world is still as dynamic as a busy hen house.

When I stand in my garden and look up at the heavens, I know that there is a unity of all things. Ours is a fruitful universe, as predictable as any Newton ever dreamed of, as unified as Einstein ever hoped for, and as dynamic as anyone dare think.

THE KEY UNLOCKS THE MYSTERY OF CANCER

There is no mystery left to cancer once it is seen for what it is — an energy mass. When normal cells are subjected to abnormal energy they run the risk of being thrown off balance. In other words, they become cancerous. When the energy is sufficient to disrupt a cell on the molecular level, a chain reaction is produced which, if left unchecked, can eventually spread through the entire body.

This is clearly illustrated by one of the best known cancer causes — radiation. Even brief exposure to this potent form of energy has produced tumors of every kind. Occasionally, there seems to be no direct contact between the source of radiation and the patient with the cancer. But in the case of painters using radioactive paint for watch dials, the connection was obvious. The factory workers used to keep their brushes moist by putting them in their mouths. There is no surprise that an alarming rate of lip and mouth cancer was evident in those workers. Had they realized that every touch of the brush was an invitation to cancer, few would have remained at these jobs.

This is only one of the scores of proven cases that can be linked to radiation. Another less conspicuous, but nonetheless deadly, cancer promoter is ordinary heat. Increased temperature means increased energy and given enough time even a fever can be as dangerous as radioactivity. Below are just a few examples of the results "ordinary" heat can have after prolonged exposure.

- Shepherds in Nepal wrap hot stones in the front of their clothes to keep warm in the bitter cold. There is a high incidence of cancer of the stomach.
- In China, the workers who dismantle the bricks of hot kilns get tumors on their forearms and wrists.
- "Sun Worshippers" constantly exposed to the sun's rays quite often develop skin cancer. (Both heat and radiation could be responsible in this case.)

The heat does not need to be from an external source. Researchers have long tried to find the link between viruses and tumors. As of yet, the only conclusion they could draw was that in some instances certain viruses are common in cancer patients. This is hardly a concrete connection. I believe that viruses do play a role, but not in the manner scientists previously thought.

It is easy for any mother to recognize when her child is suffering from a virus because that child usually has a fever. Viruses raise the body temperature. And if the virus acts long enough it could disrupt normal cell functions. That is why researchers have often found the presence of viruses in tumors. Wherever the virus is localized, the risk will be greatest.

Some may argue that if heat is indeed the culprit, then people in warmer climates should have a higher incidence of all types of cancer. This is not the case because the body is equipped to deal with the average temperature around it.

For instance, someone living in Florida will have his entire body subjected to the hot air, and it will respond by perspiring to keep the body temperature normal. It is only when a limited area of the body is being subjected to higher temperatures that the body is in danger. If heat is being applied to one area, such as a hand or foot, the body cannot use its cooling system to its capacity or it would lower the normal body temperature. As a result, those who are in contact with a constant source of localized heat, like the shepherds in Nepal, run a great risk of localized cancer.

The life span of a cell dividing varies the same as particles. But, like the healing process of cuts and broken bones, division can be speeded up when exposed to energy. When you have a hot spot, as in cancer, you have a rapid dividing of these cells.

Cancer would not show up as a hot spot unless it had excess energy. The proven cancer-causing agencies are: radiation, X-ray, heat, sun rays and irritation (which is always hot at the site). When one views this list, energy has to be deemed the culprit. Unfortunately, energy has to be spent or used, since it cannot be destroyed!

If this indeed is the nature of cancer, how should it be treated? Perhaps the best way to treat an abnormal energy mass is to try to drain the energy from it. What comes to mind is something similar to a potato battery. Many novelty catalogs sell a set of wires attached to a bulb. When the wires are inserted into a potato or carrot, the bulb lights up. If the same type of device could be inserted in a large tumor, it might be sufficient to relieve the imbalance of energy and return the cells to their normal state.

Every year the medical world hopes for the big cancer breakthrough, and every year thousands of people fall victim to this hideous disease. It is about time we treated cancer as the logical result of a simple rule of imbalance. I have no doubt that, by viewing cancer in this manner, a safe and painless treatment will be discovered.

Cancer is but one mystery that *The Key to the Universe* unlocks. The puzzle of the creation of the universe can also be pieced together using the Law of Imbalance. The world of the microscopic is the same as the world on a galactic scale. No lock is too big or too small for the key.

THE BIRTH OF THE UNIVERSE

In the beginning particles swirled through a cold and sterile space. Like cosmic bumper cars, they came together in a violent series of collisions. With each encounter, the binding forces of nature grew stronger. Electromagnetic, gravitational and nuclear forces began to pull the nebulous matter closer and closer together. Every corner of space felt the tug and every particle rushed to join the center of activity.

Due to the increasing intimacy of the pre-cosmic matter, temperatures rose, creating pockets of warmth in an otherwise numbing space. These differences in temperature served to speed up the motion of the particles and intensify the contact between them. At the heart of this coalescing mass it became hot enought to start the gases glowing. The excited electrons danced in anticipation of things to come. And what was about to come was a crucial event in the embryonic universe.

That event was the formation of the immense nuclear furnace which created the building blocks of the cosmos. When the critical temperature was reached, heavier elements with correspondingly higher energies were formed. The core now burned brightly, not only with visible light, but radiation of all wave lengths. All that we know to exist today was part of that seething mass of opposing forces. Pressure pushed outward, gravity pulled inward, and with each passing second the magnitude of each force multiplied. The scales of stability and balance were tipped beyond their limit and the point of no return was reached.

At this moment, the cosmos went into labor. Here was the first cell, filled to bursting with the power of creation, expansion and life. The encompassing placental forces were too weak to contain the energetic child. The internal forces had triumphed and the universe was born.

This birth, however, was not accompanied by the enormous "big bang" that has become so popular an explanation for the start of creation. It didn't shoot out all of its matter, scattering it haphazardly into the far reaches of space. Instead, it simply divided as any normal cell would, leaving two separate masses. For a time there was stability. But, as the forces again began to grow in strength, there were additional divisions. The "big bang" then was in reality a series of very little bangs, and the resulting universe continued to evolve in an orderly manner. We are descendants of that first cell, the product of generation after generation of these multiplications.

From the moment that first cell gave birth, we embarked on what often appears to be a chaotic journey to nowhere. Yet despite the tumultuous surface, there is an underlying reality of logic and order. Nothing occurs due to mere chance. The riddles of the universe are only riddles as long as we try to explain them by isolating them from their surroundings. We must always keep in mind that we are subject to, not exempt from, all the laws of nature. By tracing the family tree of the universe, it will become clear just how closely related we all are.

When the first egg of the universe had hatched, the

course was set for the galactic families to form. The original mass released enormous portions of matter which then began to divide on their own. The new universe continued to expand, but the majority of the matter stayed close to its immediate source. These smaller galactic-sized bundles retained the character of local families. Isolated galaxies, then, would be the exception rather than the rule.

Recent discoveries have in fact shown this to be the case. Astronomers are in the process of mapping these local clusters of galaxies and have found that they are so closely knit that they have not had the time to travel very far from the spot where they were formed. Some astronomers feel that it is simply strong gravitational forces which bind these galaxies together. However, if one discounts the "big bang" theory of violently expelled matter, and instead applies the idea of orderly reproduction, then the existence of galactic clusters is perfectly normal.

This can be illustrated very nicely by using bacteria. If a culture plate is exposed to the air, a few bacteria will land on the surface and begin to multiply. They will divide around the original spot of contact and form a very distinct area of growth. Given time in an incubator, the plate will eventually be completely covered by the bacteria. Of course, it is only after many generations that this coverage will occur. This "localized reproduction" is as common an event in the macroscopic world as it is in the microscopic world. The only difference on the scale of the universe is that it was able to grow in all directions in the limitless culture plate of space.

As these miniature replicas of the universe left their mothers' sides, their outer regions cooled rapidly and expanded in the cold of space. But while the fringes lost their heat, the cores still retained their super-heated matter.

Here, again, were the contrary forces necessary for creation. In reactions mimicking the first birth, the internal structures of the galaxies were born. As the galaxies circle the center of the universe, so the planetary systems and nebulae circle the heart of their parent galaxy. The process is like a mirror reflected into itself, ever smaller, yet always the same pattern.

It is only natural that the center of our galaxy has a black hole, for it is from there that everything in the Milky Way had its beginning. The original chunk of our galaxy has lost its ability to reproduce. But its offspring are still quite energetic and active. New centers of reproduction are being discovered all across our galaxy, affording us an excellent opportunity to witness the Law of Imbalance at work.

It is within these internal structures of our galaxy that we can most clearly see the cycle of birth and death. Our own solar system is one such structure. We need only to look up at the sky to discover its pattern. Much has been learned about our solar family in the last decade, but there is still no undisputed theory to explain how it was formed. Our system is not unlike the bacteria on the plate or the larger galaxy from which it was formed. Our little section of matter retained enough of the characteristics of its mother to start a family of its own. We are one of the most recent additions to that family and probably the best equipped to study its history.

For those bodies that are capable of reproduction there are two stages — moon and planet. When a birth occurs, the newly formed infant is a moon. As it matures, expands and releases its own satellite, it becomes a planet. We are a moon of the planet we call our sun. In turn we became a planet when we delivered our moon. Not all stars can reproduce, hence the term "planet" will be used to describe those that do.

The birth of a moon occurs for the same reason the universe was created — an imbalance of energies. A matured planet is energetic enough to release pressure in moon-sized packets — and in many cases will do it several times. Our system is the result of nine successful births from the original mass and, unfortunately, one miscarriage. The manner in which I believe this system formed may not conform to conventional theories, but it follows the same pattern as the rest of the universe and may help to solve some old questions.

The first born was Pluto. It is a tiny planet which traveled far before settling in its orbit. When the mass which was to be Pluto was first ejected, it created a shock wave like a pebble dropped in a bucket of water. The pattern of this shock wave became its orbit. For a while, Pluto orbited close to its mother's side undisturbed. It enjoyed the privileges of an only child. But the tranquility was short-lived.

Pluto soon had a sister. Neptune's mass is a thousand times greater than that of Pluto, and as it emerged it unceremoniously kicked its older sister deeper into space. In turn, Uranus pushed both Pluto and Neptune out further. With each successive birth the system expanded. Next came Saturn and then Jupiter. The size and composition of the new moons had changed drastically, which makes sense if the formation of our system progressed in this gradual and orderly fashion.

By examining the quantity and type of matter that was expelled, we can trace the magnitude of the internal struggles of the parent. Pluto came at the start of the struggle and thus was just a hint of bigger things to come. As pressure grew, the size of the discharges also increased to alleviate the imbalance. Jupiter came at the height of the struggle.

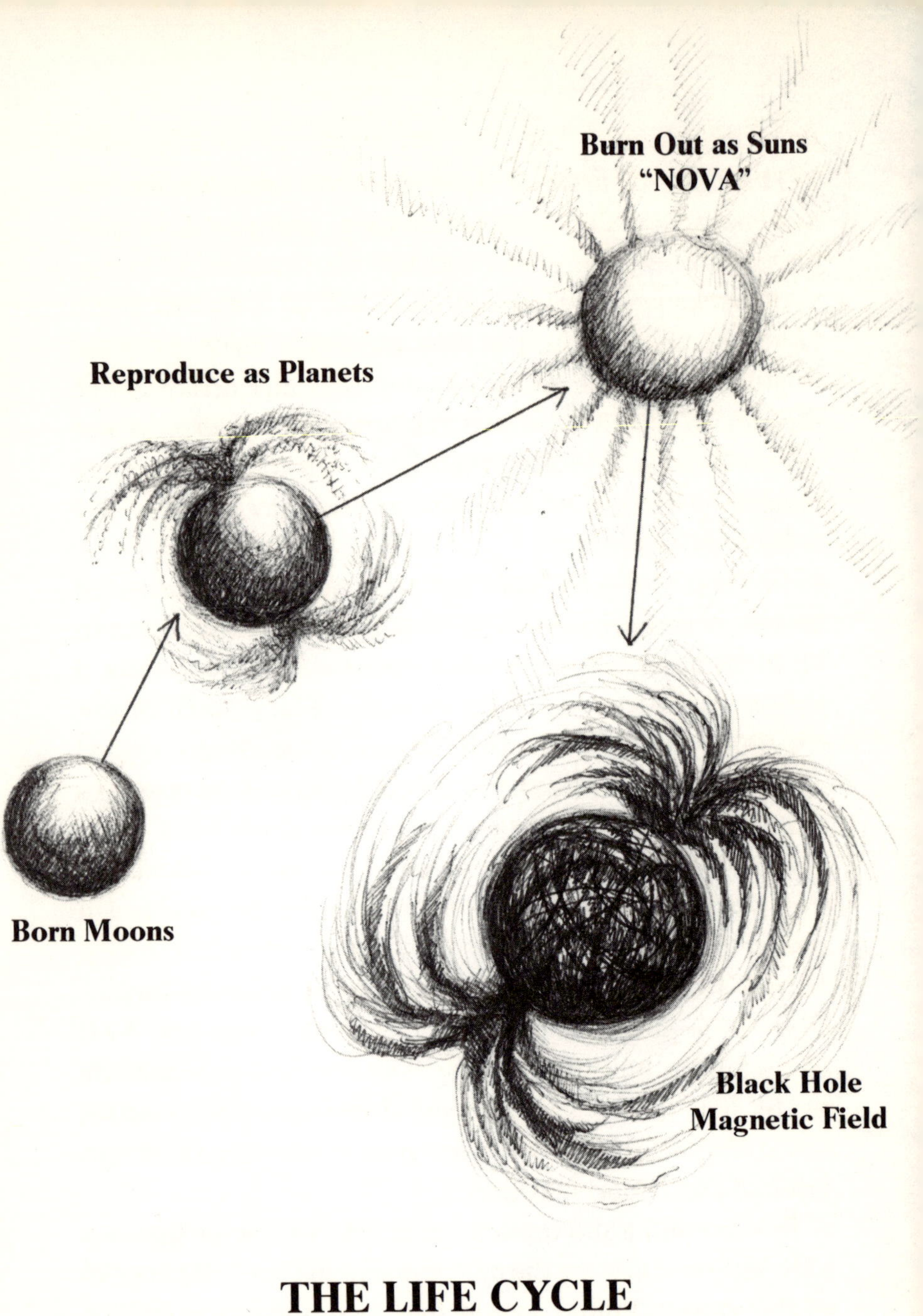

THE LIFE CYCLE
OF A GALAXY CELL

As pressure subsided, the size of the moons became correspondingly smaller. Thus a system becomes its own history book, with diagrams clearly illustrating its past events.

Few histories are without tragedy, and ours is no exception. The most catastrophic event of our early system occurred because of Jupiter's great size. Jupiter is a massive planet with an equatorial radius eleven times that of Earth and a mass three hundred and seventeen times greater. It is definitely an object to be reckoned with — as the sixth child of our system was to find out. Until this point, there had been no trouble expanding the system. Each new birth was sufficient in mass to push the others out further and establish its own orbit. However, when the sixth mass was ejected and tried to find its place among its siblings, it was obliterated like an egg thrown against a brick wall.

The mass of the infant was no competition for enormous Jupiter. While being propelled helplessly forward, it could not budge Jupiter out any farther. Caught between these two crushing forces, the young hot mass crumbled into countless pieces. Its remnants still circle in their intended orbit in the area we know as the asteriod belt. This planetary miscarriage was inevitable and would have repeated itself had anyone else challenged Jupiter's superiority. Luckily for us, the births to follow were of comparatively smaller mass than the fateful sixth and posed no threat to Jupiter or, more importantly, themselves.

When we discover other systems, they will probably follow a similar pattern. Asteroid belts will mark the spot of an unfortunate moon's demise and there will be a variety of lunar sizes. Another characteristic other systems might share is the presence of secondary moons; that is, bodies such as our moon. At last count there were thirty-three secondary bodies in our system. They are the result of seven members of our family becoming parents in their own right.

From the Earth to Pluto, the cycle of birth was perpetuated when each of them became planets. Only Mercury and Venus are still in the lunar stage. A moon becomes a planet when it releases matter in order to release pressure. This usually happens when the crust can no longer contain the molten core. The crust can, however, be gaseous, as in the case of Jupiter and Saturn, or it can be solid like Earth's.

When in the solid state, the cooling and contracting crust squeezes the center like a girdle pressing on an overweight waist. As is the case with a person, a planet can't hold its breath forever. Something has to give, and eventually it is the crust-girdle that bursts open to relieve the pressure. Matter is expelled in amounts proportional to the level of internal pressures. Just as the sun had us, our galaxy had our sun and the universe had our galaxy. So too did we give birth to our moon because of a highly energetic imbalance.

The results of a birth from a solid body can be cataclysmic, for it may be supporting a delicate balance of lifeforms. However, a body cannot resist the laws of nature. Regardless of the damage sustained by the mother, a new child is born. It will fall into an orbit in the same manner as its mother fell into her orbit. There the new moon will begin to cool and, depending upon its size and composition, it may cool completely. If the moon is volatile enough, it too could become an active member of the family.

The current theories of lunar formation are still debated issues. The scientific community has shied away from the moon-out-of-planet theory which was first put forth by George Darwin, son of Charles. George, a famous scientist in his own right, was an astronomer who extensively studied the relationships of the Earth, moon and sun. By experimenting with the stability of rotating fluids, he correctly surmised that the Earth was pear-shaped. The theory for which he is best know is his theory of how our moon was formed.

George Darwin believed that the moon came about by the action of the sun's gravity pulling matter out of the Earth before it had cooled enough to form a crust. The two molten bodies then cooled at the same time. Darwin's theory came at a time when science believed that moons and planets were basically static objects. His theory was at first acceptable because it did not have Earth playing an active role in the formation of the moon. Today, modern celestial mechanics feel that Darwin's theory is an unlikely possibility. The possibility that the Earth alone is responsible for the moon would probably be looked at as being even more unlikely. One need only to look at Jupiter's moon, Io, to see that scientists don't always have all the answers.

Io is a heaving mass of sulfurous, volcanic activity. Its fragile crust is constantly torn as the incredible strength of Jupiter's gravitational field combines with the outward push of Io's molten core. For many years, Io was just another of our system's enigmas. Observers simply could not explain why such high levels of energy could be detected on such a passive object as a moon. No one even considered that such a body could be active like the Earth. It appears that even in this day and age we like to believe that all the aspects of Earth are sacred. We only grudgingly admit that other moons and planets may be as dynamic, or more dynamic, than our own planet.

It wasn't until Voyager I passed the Galilean satellites

that the truth was discovered about Io. Several close-up photos clearly showed bright spots on the surface and arching plumes rising from the rim. The previously unnoticed phenomena were finally recognized by a NASA engineer. It was quickly revealed that Io was a volcanic moon and everyone patted themselves on the back for their startling discovery.

Had the members of NASA understood the true mechanics of planets and moons, Io would have been classified as a volcanic body long before Voyager I left Earth. Now that it is known that Io is so volatile, we should be observing it very closely in hopes of witnessing a lunar birth. All the necessary conditions are present. It would be one of the greatest moments in scientific history if we were fortunate enough to view such a birth. Hopefully, the scientific community would then realize that our solar system and its inhabitants are not exempt from the universal force of change.

In order to enlighten a member of the scientific community, I wrote him a letter in 1968. The news media had reported that there was unexplained radiation being emitted from Jupiter — unexplained only if you subscribe to the conventional idea of a dormant system, that is. I recognized it as the sign of a recent birth. I sent a letter to this prominent astronomer telling him that Jupiter must have a thirteenth moon. He replied, stating that he was not familiar with the current research on Jupiter, and that was the end of that. In 1974, Leda, the thirteenth moon of Jupiter, was discovered. I was somewhat less than surprised.

I still have a copy of my letter and the one that was sent to me. I would be interested to know if, after the discovery of Leda, the gentleman remembered the letter he had gotten six years earlier. I'm sure I will never know, but it is amusing to think that my observations might have raised an eyebrow or two.

One of the many inexplicable phenomena of our system involves the orbital perturbations experienced by the planet Neptune. There appear to be forces acting upon the orbit of the planet which cannot be explained by the other bodies that are known to exist. There are theories that undiscovered planets circle beyond Pluto, thereby accounting for the missing force. It does seem likely that there is another body of at least lunar proportions in the neighborhood of Pluto and Neptune. But the mystery doesn't end there.

Due to the unusual inclination of Pluto's orbit, Neptune periodically becomes the farthest planet from our sun. These are the only two planets in our system that go through this unusual behavior. While it could be argued that there isn't anything wrong with Pluto's orbit, it is curious that this orbital see-saw just happens to be in the same area where another member of our system is suspected of hiding. Perhaps another moon or planet would explain both phenomena. However, I do not believe that the answer lies beyond Pluto. We may have to look only as far as Charon for an answer.

There had to be a large force, such as the birth of a moon, to force a change such as this. There may not be any more information on Pluto and Charon, but I hope someone will begin to take a closer look. These kinds of disturbances are most likely to occur in the regions of Pluto where the "glue" of the system isn't as strong. The more distant planets have also had more time to become grandmothers. If Voyager I someday sends back pictures of a moon circling Charon, I'm sure the scientists at NASA will once again pat themselves on the back for a job well done.

I do not mean to condemn science as a whole. Far from it. It is simply that too many people have missed the boat. The knowledgable people who are interested in such things as astronomy have gone so far out on a limb that they have lost sight of the tree. This is, in part, understandable when someone concentrates on such a limited area of study. However, to be completely divorced from everything else should not be tolerated. It is often difficult to imagine the same forces at work in the universe around me and in my garden until I remember one thing.

Just as the vast universe grew out of one blob of matter, so too did my garden grown from a handful of simple and unassuming seeds. Planets are the seeds of the universe, driven by a common force. Our planet Earth is a very special seed. It maintains an exotic garden of countless numbers of life forms. Its past, present and future are all ruled by the Law of Imbalance. As we will see, this is a very obvious rule here on Earth.

THE SEVENTH PLANET

When Earth established itself as the seventh planet, it was a vastly different world than the one we know today. It was a smaller, denser and hotter place in its early days. The crust cooled. But the core was still super-heated and pushed out on the new crust. There are many artists' renditions of how the land mass broke apart and slowly migrated across the surface to the positions the continents hold today. Given the nature of heated objects, however, this was probably not how the continents were actually formed.

There is no logical reason for a single land mass to suddenly break apart and move in many different directions if one considers that the Earth remained constant in size. But if the Earth was once much smaller and then expanded, the present situation makes a lot more sense. The core of the Earth still maintains very high temperatures (approximately 7,800°F), and four billion years ago it must have been many times more energetic. It is common for heated objects to expand, and an object in space, like Earth, should be no different.

The outer shell of our planet cooled sufficiently to form a solid crust covering the entire surface. Volcanic activity and earthquakes built the land up, high above the molten layers beneath it. But the force of expansion was

EARTH'S 4 EVENTS

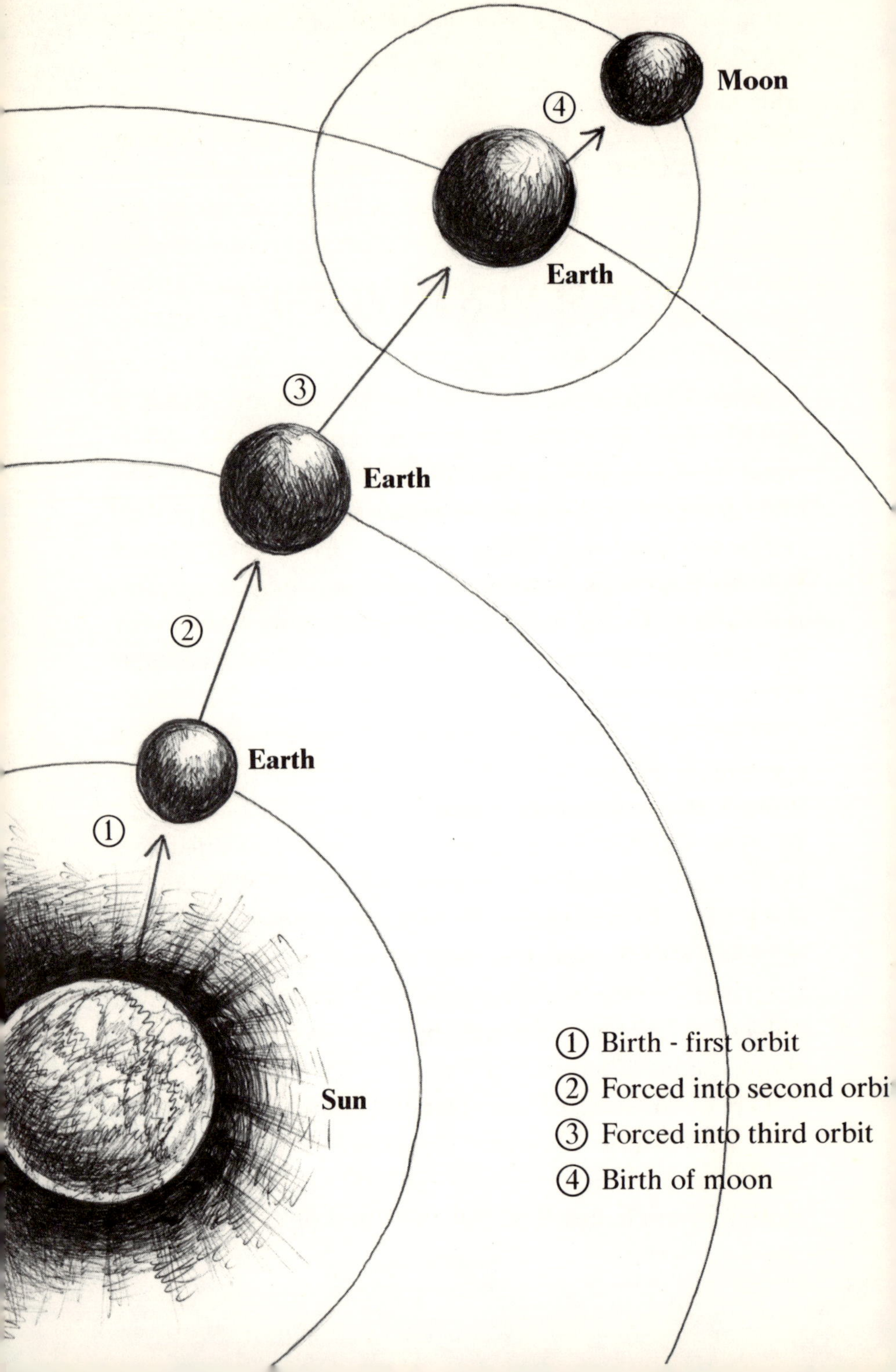

not to be stopped, and the outer shell split into pieces. As the globe expanded, molten material rushed upward. The gaps between the continents were filled, but the original outlines were not erased. The newer, lower crust became the ocean basins, while the higher land remained dry.

A balloon and some mud serve well to demonstrate this idea. If you put a thin layer of mud on a partially inflated balloon and allow it to dry, you have a model of the Earth in the early stages. When the balloon is inflated to its full size, the layer of mud will break apart and move along the expanding surface. Cracks, resembling faults on Earth, will also streak the surface of the mud. Wherever the pressure is the greatest is where the large cracks will appear. Of course there were many other forces at work than are illustrated by the balloon, but the basic idea of an expanding world is the same.

The Earth is still in the process of being inflated as the molten layers push upward on its all-too-fragile crust. Anyone who has witnessed the eruption of a volcano or lived through an earthquake is well aware that our planet is far from being a dormant body. In fact, there is an earthquake on the average of every thirty seconds. When George Darwin made his conclusions about the sun pulling matter from the Earth, he was wrong. But he was right in assuming that the matter for the moon was from the Earth. He simply underestimated the power of our own planet.

There have been many ice ages throughout history — proof that the Earth does go through periods of cooling down. When a handful of volcanoes isn't sufficient to accomplish the required cooling, something has to give — namely the crust. The intensity of the pressure must have been great indeed when the moon was born, for the amount of material that was expelled is substantial compared to the size of the other moons in our system. The birth of our moon was also devastating to this planet.

There are many indications that the Earth was already teeming with life when it became a mother. The Bible speaks of the time before Earth's moon was born. It tells of a great flood, as do legends of India, South America and numerous other civilizations. Other legends from around the globe say that the sun used to rise in the west before a terrible catastrophe occurred. Fossil beds of tropical plants and animals have recently been found in the area of the North Pole. There are too many signs to ignore or explain away.

The shock that would have accompanied the birth of the moon would have been strong enough to shift the Earth's axis. This would account for the legends of horrible winds and earthquakes wreaking havoc before settling down, and of the sun beginning to rise from a new direction. If the Earth did turn over, the old poles were probably located in the Bermuda Triangle and the South China Sea. Perhaps all the strange magnetic disturbances of the Bermuda Triangle are the remnants of the magnetic field of the old poles.

Even if suitable answers can someday be found to explain those puzzling legends and tropical fossil beds, there is still the problem of the dinosaurs. Scientists have blamed the extinction of these huge reptiles on everything from meteors to constipation. Whatever the cause, the results were fairly swift and very thorough. Even if a few dinosaurs had survived the initial trauma of the lunar birth, the ensuing cooling down period would have finished them off. This is not an unreasonable theory, for life on Earth is certainly not immune to catastrophes. Just ask any dinosaur.

The radiation which must have accompanied the birth of our moon would also have been quite deadly. Any creature that survived would have been practically free from competition. However, the radiation's effect on the future generations would be one of mutation and possible sterility. Was the missing link actually a radiation bath? There is no doubt that if the moon sprang from the Earth, life on this planet would be vastly different afterwards.

One argument against the moon coming from the Earth in comparatively recent times is the age of the rocks tested by the nine lunar samplings (six U.S. and three Soviet). Some estimates claim that the youngest rocks were formed approximately 3.1 billion years ago. However, it is possible that the dating techniques used here on Earth are not valid on bodies lacking our thick atmosphere. Cosmic rays and high energy atoms from the solar winds have bombarded the surface of the moon since its birth. The steady stream has continued unchecked by an atmosphere. Just as a car that is kept in a garage looks newer than one left at the mercy of the elements, so the moon has spent its lifetime without the protection of the Earth's garage. In the future, more accurate methods of determining age may hinge on the ability to gauge the effects of the cosmic elements.

Perhaps more can be learned about the effects of a lunar birth on its parent planet by studying Mars. Mars has two moons, as well as the largest volcanoes in the solar system. There are also polar ice caps containing the water which must have flowed freely across the surface at one time. Mars now appears to be in a dormant stage. However, when the water wasn't frozen and the planet was warmer, there is a good chance that there was life. Tests performed by the Viking landers failed to find any signs of life — but some scientists feel that the place to look is by the water.

However, because a planet does not support any life now doesn't mean that in its long history it was always barren. If the effects of a lunar birth were that severe, all traces of life could have silently disappeared under the layers of red dust and hardened lava. If there was an advanced civilization on Mars, perhaps the inhabitants picked up and left. Whatever the situation on Mars was in the past, we may be able to benefit from it in the future.

There are many popular theories about the Earth being visited by beings from other planets. Perhaps people from Mars, realizing their planet was doomed, did what any Earthling would do — run to the nearest and safest spot. Earth would have looked awfully inviting to anyone who was about to have his civilization wiped out. Maybe it was these Martian refugees who gave rise to the ancient legends of visitors from the sky. They might have helped to build the pyramids or etch the strange markings in the plains of Nazca.

If the beings of Mars didn't leave in time to avoid the radiation and frigid temperatures, they may at least have left messages for future travelers. The message probably would have described just what happened when their world suddenly came apart. We may want to take a minute and reflect on the plight of these hypothetical Martians, for we also live on a planet that is subject to radical change. And like them, we too may someday need a place to run to.

LIFE BEGINS

Primordial soup has been on the menu for most scientists in recent years. It is believed that life arose out of a warm, murky mixture of methane, ammonia and water. The primitive Earth was abundant in these substances and there was plenty of lightning to get things going. Several laboratories have duplicated this soup. The necessary ingredients are placed in a large glass bowl and subjected to artificial lightning. Amino acids, the building blocks of life, are formed in this reaction.

No one has yet succeeded in creating life from these experiments. But they have demonstrated that the necessary building blocks could have been formed. Perhaps with a pinch of this and a dash of that, life could begin anew in a piece of laboratory glassware. For now, however, we will have to be content with thinking we are on the right track. However, not everyone feels that this is the answer.

There are those who believe that life on Earth was seeded from an extraterrestrial source. Theories range from cells flying in on meteors to cells being flown in by people. These same people are often credited with developing the human race by selective breeding. Even if this was the case, the people who brought life to Earth had to begin somewhere. Which brings us back to square one. Where life originated is not the important thing. What is important is that life is here and we have the power to discover its secrets.

Life is a tricky thing to define. About the best determination that can be made is to say that what is able to grow, reproduce and respond to stimulation is alive. That may or may not be the definitive answer, but it certainly doesn't even hint at how life began. If the primordial soup theory is correct, life in its infancy was no more than a group of molecules that had somehow divorced themselves from the equilibrium of the surrounding soup. The most crucial test that was passed on was the ability to maintain balance versus imbalance.

If these few molecules were able to protect themselves with a membrane shield, they would have been able to increase the differences between themselves and the rest of the mixture. By simple diffusion, the needed chemicals could have been passed through the membrane and the unwanted ones passed out. This is assuming, of course, that the primitive cell had some control over what came in and out, as well as "knowing" what it needed. As pockets of independent molecules grew in size, areas of the pocket became specialized. Perhaps this specialization arose by the joining of different cells. Whatever the case might be, the point is that in order to begin life, the outside mixture had to be different than the internal mixture.

The simplest forms of life are barely more than membranes which can maintain a concentration gradient. Ion exchange across a membrane is the driving force behind all metabolic functions. This exchange is essential for everything from a single-celled animal to a human brain cell. If the imbalance between the cell and its surroundings is impaired or stopped, the cell dies. So, like the universe, man contains a multitude of opposing forces crucial to his existence.

Another similarity to the universe is the fact that the internal structures became highly differentiated. There are mitochondria to generate energy, ribosomes for protein assembly, a nucleus to store genetic material and numerous other structures that play their own specialized role. Each cell of the body is, in actuality, a living organism in its own right, with the ability to grow and reproduce. However, like a planet, a cell will not divide unless provoked by some form of energy.

In the case of animals, hormones trigger the cells. Hormones co-ordinate and stimulate (or suppress) body functions. They are the chemical "imbalancers" that signal cells and organs to excrete, reproduce or do nothing. In much the same way that radiation sparks cancerous cells, the body sparks itself. Many problems can arise if the body loses control over its energetic hormones. Because a cell only reacts to the amount of stimulus it receives, it won't stop functioning regardless of the harm it may be causing to the organism of which it is part. Thus hormone imbalances can cause dwarfism, giantism and a range of diseases from diabetes to goiter.

It is clear by now that all energy, whether chemical, thermal or radioactive, must be respected for its potential hazard and recognized for its critical role in our lives. There is a force common to all things. It is the force created when any type of energy tips the scales of balance. Whether you hatch an egg, grow a tumor or create a solar system, you are employing the universal force of imbalance.

THE EARTH GROWS UP

The Earth supports an intricate chain of life. It has created and nurtured it in her green forests and blue oceans. Even areas as harsh as the desert and as freezing as a snow field, maintain a multitude of plants and animals. However, there are times when the Earth becomes so cruel that not even the hardiest forms of life have a fighting chance.

Volcanoes, floods and blizzards may be deadly to living things, but we must remember that these things are necessary for our living Earth. Changes in the weather do not occur in a random manner. There are very concrete reasons why the wind blows, the rain falls and the temperature rises. The shift of these conditions directy reflects the influences at work both beneath the surface and beyond the atmosphere of our planet.

If all volcanic and earthquake activity was stopped for even a short time, the Earth would probably blow up from the pressure. We are very much like a civilization living on the lid of a pressure cooker. As destructive as a quake or eruption might be, the alternative is much worse. As this activity increases and decreases, the weather varies correspondingly. Thus a period of heavy activity would precede an ice age.

There have been at least four major ice ages in the

Earth's history, each lasting several million years. The last ice age began roughly 1.9 million years ago; with the last retreat of the glaciers beginning as little as twenty-five thousand years ago. These extended periods of cold and then hot reflect the Earth's growing pains. After enough pressure and energy was released from the mantle, temperatures would drop. As the Earth began to get active again, the glaciers receded and melted, the oceans rose, and the surface was again warm and moist.

The scarred surface of the Earth is a vivid reminder of the pushing and pulling that the evolving planet experienced. Fault lines, reminiscent of stretch marks, streak the globe. Mountains thrust miles above sea level stand as a testament to the herculean forces at work beneath us. Islands literally are built and destroyed overnight. What is a major catastrophe to us is merely a typical day in the life of our dynamic Earth.

Life is indeed a tricky thing to define. For by our definition of growing, responding to stimuli and reproducing, our Earth is very much alive. Perhaps in the light of a universal force, there is little distinction to be made between living and non-living. Mother Earth has proven herself to be fertile and well able to regulate the vast amounts of energy at her disposal.

When the young Earth matured, it gave birth to the moon. As we have seen, our host is far from being a retired old lady. This vitality could put us in a perilous position. But before that possibility is discussed, the fate of the rest of the universe deserves our attention.

UNIVERSAL AGING

Nothing is immune to the universal force of change. Part of that change involves aging, and the planets and moons have their own way of experiencing this change. The best way to describe the process is to use the oldest object as an example. In the case of our universe, the oldest object can be found at the center, the spot where it all began.

The mass at the center of the universe divided many times, sending the beginnings of galaxies out into the far reaches of space. While astronomical proportions of matter left the area, not all of it was dispersed. What remained was still a massive amount. But as the original powers of creation began to weaken, the matter began to go through many changes. As the internal forces lost their potency, the outer shells of material started their collapse inward. This theory of collapse has already been put forth to explain the proposed black holes, but it is usually limited to only certain stars. It is a valid theory. However, its applications are far more widespread than previously thought.

The remnants of the original mass did begin its journey toward becoming a black hole, but not before it put on one last display of might. As the pressure of the collapsing material stirred up the failing forces of the core, new reactions became possible. Material which had been unable to undergo nuclear reactions in the lower pressure and temperature of the original mass sprang to life in the new increased pressure. The core now had the strength to thrust the outer shell away and expand once again to gigantic proportions.

These last acts of defiance have been visible to Earth many times as previously faint specks of light turn into novas. However, these bodies are burning on borrowed time as the limited supply of their fuel is soon depleted. The swan song of these once fertile bodies is a spectacular reminder of the powerful creative forces of their youth. The center of our universe must have been an incredible sight as it flexed its muscle a final time before beginning the inexorable collapse toward a black hole.

This time there were no untapped reserves of fuel. Matter rushed inward at ever increasing speeds as the gravitational forces grew around thc dense core. The light dimmed as the final bits of matter burned out. As gravity reached monumental proportions, even the dying flickers of light could not escape its pull. Gravity became a titanic force, but it was not the only force to increase in strength.

The magnetic field also grew in strength. In other aspects, a black hole is a ghost of its former self. But the magnetic field still exists and is a dominating feature. So, despite the lack of any more visible display, the dark center of our universe is still a very dynamic area. We have seen that the universe has followed the pattern set forth by the original mass, and the aging of a parent is no different. As the matriarch of the universe aged, so too did the centers of the galaxies, on down to the level of the solar system.

Bodies are born as moons, mature as planets, burn out as novas and die as black holes. Our galaxy was a moon of a larger mass. It matured and spawned a multitude of systems, clusters and nebulae. As the center went through the changes that the core of our universe experienced, it too burned brighter than ever. Now it is in the black hole stage. It is a black hole enormous enough to still influence the entire galaxy. Every productive galaxy will eventually have a supermagnetic black hole in the site of its origin.

Some evidence has already been discovered which may reveal that our galaxy does indeed have the same awesome force at its center as does our universe. Plumes of gas have been detected shooting out from the center of our galaxy in long arching jets. This super-heated ionized gas may be caught in the powerful magnetic field radiating from the core. Every schoolboy and girl knows the pattern that iron filings make around a magnet. They arch about the magnet in loops. These loops, called flux lines, spring from the top of the magnet and join it again on the bottom. The gas jets coming from the galactic center would seem perfectly natural if we viewed the galaxy as a huge galactic magnet. It would not be surprising if similar jets were found emanating from the other side of the center and meeting in the middle along the rim of the galactic plain.

In the future, our search for black holes should include a search for similar jets of hot gases curving outward from an area suspected of containing one of these peculiar bodies. Research is currently being conducted on several likely candidates. Such a strong possibility as Cygnus X-1 should exhibit these jets. The universe must be teeming with black holes, and a stream of gases should point the way to each of them.

Not every object in the galaxy goes through the complete cycle of moon to black hole. Many objects never get past the lunar stage. Some of the smaller moons in our system are already no more than cold chunks of rock. The fate of each body depends upon its internal conditions and external influences.

The brightness and duration of the nova, as well as the potency of the resulting black hole, also depend on the nature of the body during the earlier part of its life. Someday our system will follow the path of its elders. Each body will undergo the aging process at different times and at different rates. But undergo it they will.

PHYSICS – THEN AND NOW

If the Law of Imbalance is truly a law, it should be applicable in all areas — from the universal, to the cellular, to the sub-atomic. In order to discuss this law in relation to sub-atomic particles, it would be helpful to give a general background of physics.

Physics at the turn of the century was dominated by one man, Sir Isaac Newton. It was upon his view of a mechanical universe that all of physics was based. Newton arrived at his conclusions by elaborating on Euclidian solid geometry. By assuming that time and space were separate and that particles were immutable, he developed a theory of a universe that never changed. This was no small feat. In order to explain his theory fully, however, he needed a type of mathematics that didn't exist.

To accomplish his goals, Newton invented calculus. Einstein described this as the most incredible achievement by a single human being. Both his invention and his theory were extraordinary achievements, and they did help to explain the world as it was then known.

However, what was known in the past and what is known now are two completely different things. In all fairness, if Newton had had access to the information we now have, a brilliant mind such as his would probably have solved many of physics' problems. Since Newton wasn't available, Albert Einstein had to step in.

In 1905, Albert Einstein published two articles that would turn the field of physics upside down: one on electromagnetism; the other on relativity. All that Newton had believed, and all that physics had been for the last several hundred years, suddenly was changed. The universe of the unchanging became the universe of change.

The world of physics in this century has been one of startling revelations which seem to defy reality. Most of us are aware that atoms are composed of protons, neutrons and electrons. However, few are aware of the smaller particles with such exotic names as leptons, mesons and barytrons. It is at the level of these particles that modern physics has made the greatest and most baffling discoveries.

In order to study the world of the sub-atomic, particle accelerators, over four miles long, were invented. One is located in Batavia, Illinois, at Fermilab. Named after the famous nuclear physicist Enrico Fermi, the accelerator is a round track in which protons are accelerated to speeds approaching the speed of light. The highly energetic protons are then shot into a chamber containing particles such as other protons or neutrons. There are numerous collisions, and the resultant shower of different particles streaks across the chamber.

The tracks these particles make can be photographed. By examining the pattern of the lines, the type of particle that produced them can be identified. As we have already seen, off-balance substances reproduce. It is perfectly natural, therefore, for even a sub-atomic particle to give rise to a new particle when subjected to abnormally high levels of energy. For example, an energetic tennis ball would not break apart; it would become two tennis balls.

The time such particle reactions take are incredibly small fractions of a second. This, too, follows the pattern of the universe. On a universal scale, reproduction is accomplished in billions of years. Cells generally need only a few hours. It appears that the simpler and closer a substance is to pure energy, the faster it reacts to additional energy.

In reality, the distinction between matter and energy on the particle level is meaningless. They are little more than tiny packages of energy. Ever since Einstein wrote his formula $E=mc^2$, it has been known that matter is composed of energy. But it has never been so clearly illustrated as it is in particle accelerators. There is a very fine line between balance and imbalance at this most basic of levels.

THE PERIL OF MOTHERHOOD

There have been numerous instances in the past when our planet and its inhabitants underwent great changes due to an imbalance. The land and sea have shifted across the face of the globe, and what was once a desert or a lake may now be the opposite. Fossils of marine life have been uncovered in the Arctic and high atop mountains. In our own recorded history, civilizations have been obliterated by floods, volcanoes and earthquakes. Modern cities struggle to keep their towering structures from toppling down in regions where quakes are severe. There are constant reminders of the fact that our Earth is restless. In the light of all these facts, can we be certain that the Earth has finished bearing children?

It is difficult to write this chapter because it appears as if the Earth will not stop at one child. If one considers only the other planets in our system, the trend seems to be toward families of more than one moon.

The last few years have seen peculiar changes in our weather. Temperatures have begun to creep a few degrees above normal. At first, this was attributed to an enormous cloud of dust and ash that had been sent into the atmosphere by a volcano. This cloud stretched across the globe and could have acted like a blanket, trapping in the warm air. However, even if this cloud did affect our weather for a short time, it cannot account for all of the other shifts in temperature around the world.

There also appears to be an increase of volcanic and earthquake activity. Volcanoes such as Mount St. Helens are evidence that beneath a tranquil surface there is a violent world. It could only be a matter of time before other volcanoes in the United States and elsewhere reawaken, and with catastrophic results. If there is indeed a marked increase in activity in the near future, we may witness a lunar birth.

What would life be like if such an event were about to take place? And how would we deal with it? Just as we hypothesized about the people of Mars facing extinction, we, too, may be about to face the same situation. As the temperature of the mantle increased, the surface of the Earth would go through drastic changes long before the actual birth of a moon occurred.

One of the immediate reactions would be the melting of the polar ice caps and the subsequent rising of the oceans. An increase of only a few degrees could melt the ice and liberate an enormous amount of water from its frozen storehouse. Coastline communities and cities would soon be submerged and the populations of the world would have to move inland. The heat would drive more humidity into the air as the entire planet became a warmer and wetter place.

There is no telling exactly what disastrous effects rising temperatures would have on agriculture. It is safe to say that large areas of vegetation, as well as animal life, would be devastated. As the time of birth grew near, few things would be able to flourish in such extremes of heat and humidity.

Even if there are those species that might survive such a climate, the lack of sunlight may spell their end. The ash and debris from hundreds of volcanoes will set up a permanent cloud, blocking precious rays of sunlight. Faults will split open and new mountains will be thrust upward as pressure from below reaches critical proportions. Few places on earth will be safe from all of the inhospitable changes.

Hopefully, we will have discovered a solution before the point of no return has been reached. Whether we are even capable of altering such a course of events remains to be seen, but we would have to try. There are many experiments currently being conducted which use geothermal energy to produce electricity. Set deep into the crust, the intense heat will generate steam in the pipes and send it to a power plant on the surface. By tapping the incredible energy of the magma, enough electricity could be generated to power the entire United States. In fact, it could easily generate thousands of times the amount of electricity used in this country over the course of a year.

If we do discover that our planet is about to give birth, we may use the knowledge gained thru these experiments to find a way to release some of the heat and pressure. If enough vents can be opened, we might be able to assist the Earth in releasing the pressure. Perhaps a method such as this could be employed now to try to control the severity of earthquakes in danger zones such as California and Japan.

Plans for space stations and orbiting hotels have already put science fiction well down the road to reality. The colonization of Mars is also becoming a real possibility. As terrible as it would be to have to leave Earth, it is comforting to know that alternatives are possible. Even if conditions never reach a critical point, there may be periods when the population will want to temporarily retreat until things cool down.

It is imperative that the scientific community begin to address this problem. The time of another lunar birth may be closer than we think. Perhaps the legends and stories of future destruction will come true — not as the result of battles going on above the surface of the Earth, but beneath it. This is certainly too crucial an issue to ignore.

THE POWER OF BALANCE

Forces grow in strength and forces are diminished, but always there seems to be some force at work. Among the many forces suggested to be at work in the world is one known as Pyramid Power. This has been publicized to the point where people wear little pyramids on their heads and set up pyramids in their living rooms to watch television. So much has been said about this force that it is hard to separate fact from fiction.

What does appear to be true is that objects such as fruit and dead animals, when placed under a pyramid, do not rot like they would if they were outside of the pyramid. There are many legends about the healing powers of the Great Pyramid, but there is not very much in the way of documentation. It is certain that the Egyptians built the pyramids at Giza in a carefully planned manner and for a very specific reason. They believed that the shape and dimensions of the structure were conducive to whatever it was they were doing inside. The question is: What exactly were they doing? And how did the pyramid help?

Perhaps the answer is not Pyramid Power but "Pyramid Lack-of-Power." It may be that the structure does not create forces but rather neutralizes them. If this is the case, then a pyramid would be the perfect place to preserve as well as to heal. This could explain many of the mysteries surrounding the legends. An area free from the turbulent forces in the world would be a peaceful haven for anyone caught up in the daily struggles of life. Anyone who was ill would find it a soothing atmosphere in which to recuperate.

There is much to be said about the creative force of imbalance. However, there comes a time when balance is needed to repair or simply to maintain a situation. Like the Hindu idea of creation, sustenance and destruction, the power of balance and imbalance interact to form the universe. The status of the universe at any given moment is a direct result of the proportion of balance and imbalance.

By manipulating *The Key to the Universe,* we will be better able to control our lives here on Earth. In all areas, from medicine to space travel, the "key" can unlock a world of greater achievement and less suffering. There are unlimited possibilities in our universe, and we should waste no time in exploring them. We now have the greatest gift in the universe at our disposal — the gift of knowledge — and we must utilize it swiftly and to its fullest extent.